Cambridge Primary

T0187276

Hodder Cambridge Primary

Science

Activity Book

B

Rosemary Feasey

Foundation Stage

HODDER
EDUCATION
AN HACHETTE UK COMPANY

The author and publishers would like to thank Chris Lawson, Science and Early Years Lead, Laurel Avenue Primary School, for her support in planning this material.

Orders: please contact Hachette UK Distribution, Hely Hutchinson Centre, Milton Road, Didcot, Oxfordshire, OX11 7HH. Telephone: +44 (0)1235 827827. Email education@hachette.co.uk Lines are open from 9 a.m. to 5 p.m., Monday to Friday. You can also order through our website: www.hoddereducation.com

© Rosemary Feasey 2018

Published by Hodder Education

An Hachette UK Company

Carmelite House, 50 Victoria Embankment, London EC4Y 0DZ

Impression number   10 9 8

Year   2024

Cover © Steve Evans

Illustrations by Vian Oelofsen

Typeset in FS Albert 17 pt by Lizette Watkiss

Printed in the United Kingdom

A catalogue record for this title is available from the British Library

978 1 5104 4861 2

Contents

What makes a sound?

We use our sense of hearing to listen to sounds. Our ears hear sounds.

⭐ Draw and write four things that make a sound.

⭐ Draw things that make these sounds.

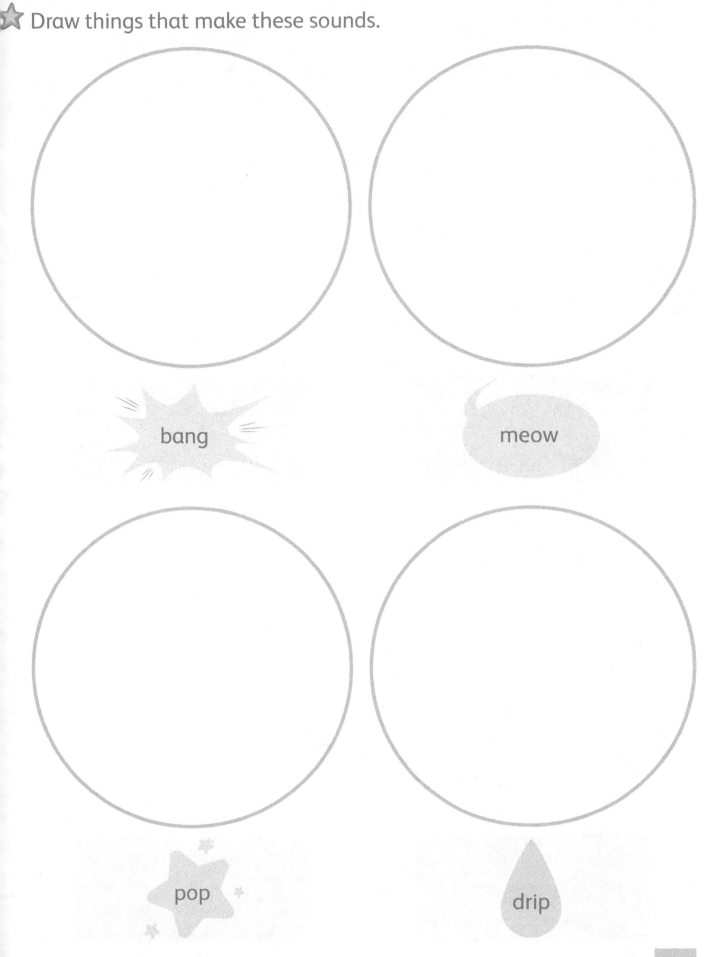

bang

meow

pop

drip

Sorting sounds

 Draw sounds you like. Draw sounds you do not like.

Sounds I like

Sounds I do not like

⭐ Which things make a sound? Join each picture to a ✔ or a ✖.

Makes a sound

✔

Does not make a sound

✖

Animal sounds

⭐ Write the sound next to each animal.

| hoot | baa | buzz | squeak |

bee

owl

mouse

sheep

Circle the sound for each animal.

| quack | tweet | baa |

| croak | roar | squeak |

| meow | growl | hiss |

 Draw a different animal that makes a sound.
What sound does it make?

Sam's sound hunt

⭐ Sam is going on a sound hunt. Circle the things that make a sound. Join the word to the sound.

| rustle | shout | vroom |

| splash | tweet | ding |

Go on a sound hunt! Draw and write the objects that make the sounds.

Musical instruments

 How do these instruments make a sound?
Join each instrument to a word.

tambourine

hit

scrape

guitar

blow

pluck

guiro

panpipes

 Look at the instruments above.
Draw and write the one that you can shake **and** hit.

Play some instruments!
Draw the instruments
and write how you
make a sound.

| pluck | hit | scrape | blow |

This is a

_____.

I _____ this instrument to make a sound.

This is a

_____.

I _____ this instrument to make a sound.

This is a

_____.

I _____ this instrument to make a sound.

Loud and soft

Sounds can be loud.

Sounds can be soft.

⭐ Make a loud sound. Make a soft sound. Draw and write what you did.

loud

I made a loud sound by _____.

soft

I made a soft sound by _____.

⭐ Write **soft** or **loud** next to each picture.

[loud]   [soft]

lion roaring _____

mouse squeaking _____

balloon popping _____

⭐ Put some rice in a box with a lid. Shake the box. Is the sound **loud** or **soft**?

The sound is _____.

⭐ Put something different into the box. Is the sound loud or soft?

I put _____ into my box.

It made a _____ sound.

Make an instrument

You can make instruments from junk materials. You could make ...

a tin can drum.

a pipe cleaner rattle.

a cup shaker.

a rubber band guitar.

⭐ Design an instrument. Draw a picture. Write what you will use.

Make your instrument. Write the name of it.

I made a _____ .

Stick a photo of your instrument here.

 Write what you do to make a sound. Use the words to help.

hit	shake	blow

pluck	loud	soft

To make a sound, I _____ my instrument.

My instrument makes a _____ sound.

 Play your instrument with your friends.

Can you make a loud sound?

Can you make a soft sound?

17

Sorting toys

Toys can be sorted in different ways.

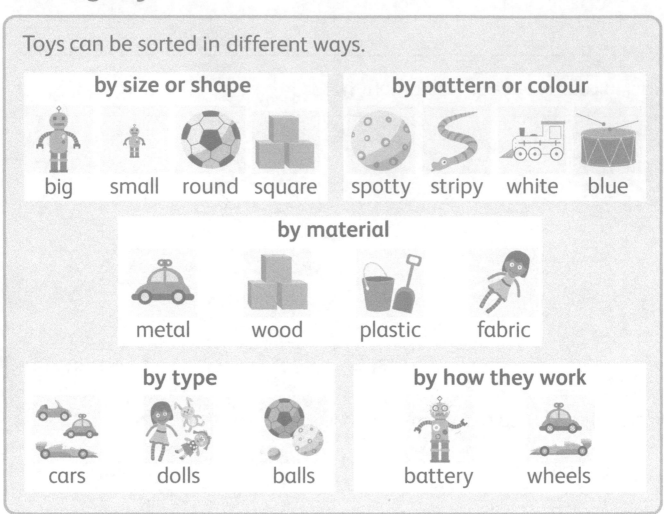

by size or shape

big   small   round   square

by pattern or colour

spotty   stripy   white   blue

by material

metal   wood   plastic   fabric

by type

cars   dolls   balls

by how they work

battery   wheels

⭐ Choose a way to sort some toys. Draw and write how you sorted the toys.

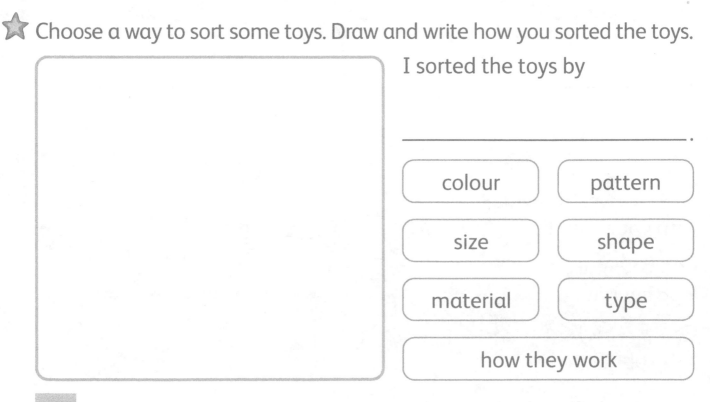

I sorted the toys by

_____.

| colour | pattern |

| size | shape |

| material | type |

| how they work |

⭐ These toys are made from different materials.
Join each toy to the correct toy box.

wood

metal

plastic

fabric

Moving toys

Toys can move in different ways.

push

pull

Pushes and pulls are forces.
Forces make things move.

⭐ How do your toys move? Draw a toy in each box.

I can push this toy.

I can pull this toy.

Choose a toy to push and pull.
Find out how the toy moves on different surfaces.
✔ to show what you find out.

I pushed my toys on …	It was easy. ☺	It was not easy. ☹
grass		
soil		
sand		
classroom floor		

Which surface was the easiest? _____
Say why you think it was the easiest.

I pulled my toys on …	It was easy. ☺	It was not easy. ☹
grass		
soil		
sand		
classroom floor		

Which surface was not the easiest? _____
Say why you think it was not the easiest.

Magnetic toys

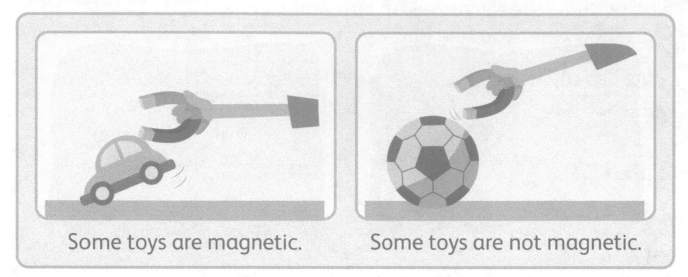

Some toys are magnetic.   Some toys are not magnetic.

⭐ Test some toys with a magnet. Draw what you find out.

These toys are magnetic.	These toys are not magnetic.

⭐ Put some things inside a bottle with a lid. Move a magnet up and down the bottle.
Which things are magnetic?

The magnetic things are made from

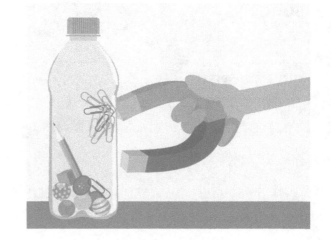

_____

Exploring play dough

⭐ Draw and write what you can do with play dough. Use these words.

stretch squash squeeze

roll twist

I can _____ play dough.

I can _____ play dough.

I can _____ play dough.

Bubbles

 Blow some bubbles.
Say how to make
big bubbles.
Say how to make
small bubbles.

 Design a bubble blower. Draw and write what you will use.

 Test your blower. Does it work? _____

Does it make small or large bubbles? _____

Shadow puppets

 Collect some toys and a torch.
Try to make shadows.

 Design a shadow puppet. Draw and write what you will use.

 Make your shadow puppet. Test it using a torch.

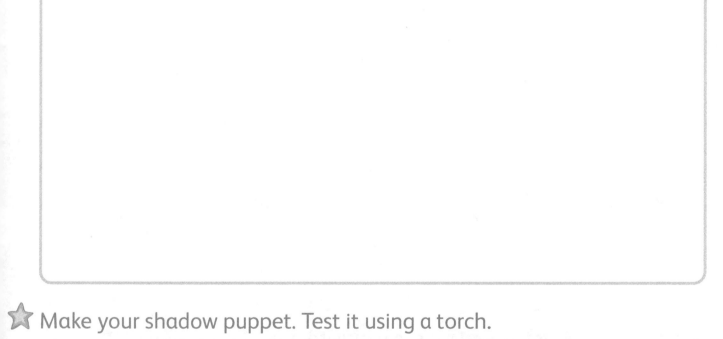

Build it!

 Build a tower.
Draw a picture of your tower.

 Write about your tower. Use the words to help.

| taller | shorter |

My tower is _____ than me.

I used _____ to build my tower.

Design a bridge for small toys.
Draw and write what you
will use.

 Test your bridge. Write what you did.

I used _____ to build my bridge.

My bridge can carry _____ .

Make a toy

You can make a toy using junk materials. You could make …

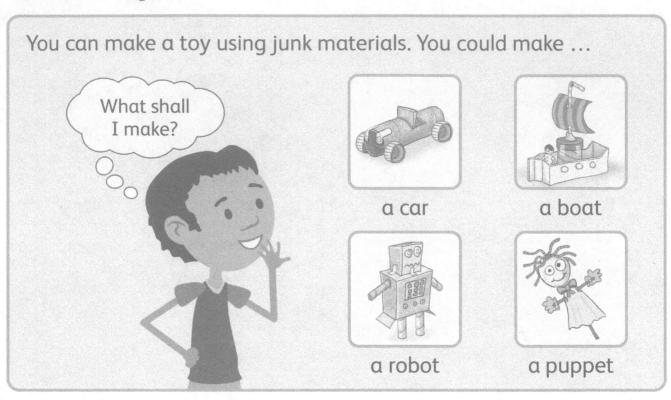

What shall I make?

a car

a boat

a robot

a puppet

⭐ Choose a toy to make. Draw your design. Write what you will use.

⭐ Make your toy. Try it out!

⭐ Write about your toy.

I made a _____ .

⭐ What is your toy made from?

(wood)   (plastic)   (cardboard)   (paper)   (fabric)   (metal)

My toy is made from _____

_____

_____ .

⭐ How do you make your toy move?

(pulling)   (pushing)

I can make my toy move by _____ it.

⭐ Take a photo of your toy.
Stick it here.

What can you remember?

⭐ Draw and write three things that make a sound.

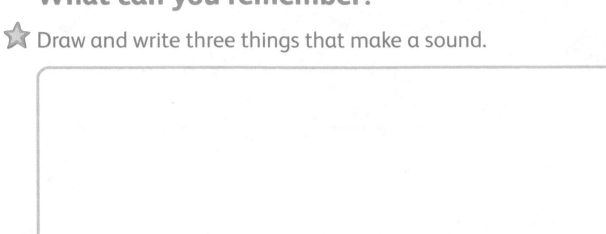

⭐ Join each instrument to how it makes a sound.

shake

pluck

blow

hit

⭐ Put a circle around the things that **do not** make a sound.

hairbrush

snake

table

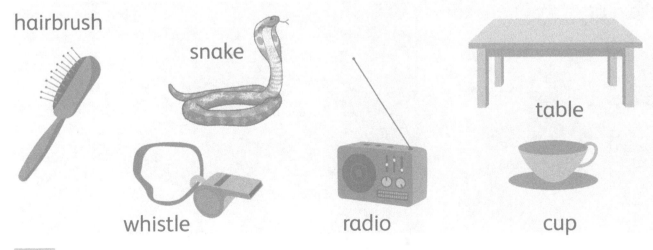

whistle

radio

cup

Which toy is the odd one out? ✓ it. Say why.

| train ☐ | doll ☐ | scooter ☐ | car ☐ |

 Join a toy to its shadow.

doll's house

car

skipping rope

ball

 Circle the toys that you can push.

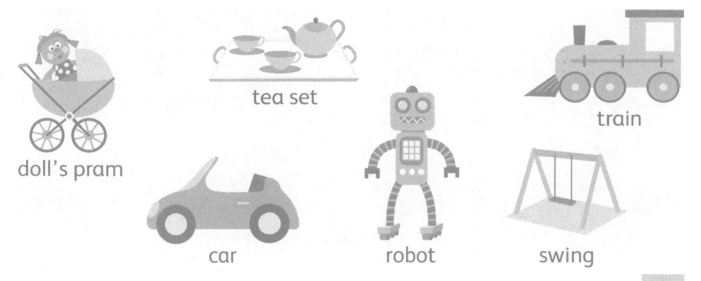

doll's pram

tea set

train

car

robot

swing

Self-assessment

Colour the stars to show what you can do!

Making sounds	I can say which things make a sound.	☆
	I can describe the sound things make.	☆
	I can sort sounds into groups.	☆
	I can say what kind of sounds animals make.	☆
	I can say how different instruments make a sound.	☆
	I can make loud and soft sounds.	☆
	I can make an instrument and describe its sound.	☆
Toys	I can sort toys into different groups.	☆
	I can name toys that move by pushing and pulling.	☆
	I can show which toys are magnetic.	☆
	I can make and describe different shapes using play dough.	☆
	I can make a shadow.	☆
	I can say how to make big and small bubbles.	☆
	I can make a toy and describe how it moves.	☆